Chapter 1: The Basics of Chemistry

Understanding Chemical Compounds

At the heart of all matter lies chemistry, the science that explores the properties, composition, and behavior of substances. Chemical compounds are formed when two or more elements combine in fixed proportions, resulting in a unique substance with distinct properties. Understanding these compounds begins with grasping the fundamental building blocks of matter: atoms.

Atoms, the smallest units of an element, consist of a nucleus surrounded by electrons. The arrangement and number of these atoms dictate the characteristics of a compound. For instance, water (H_2O) is formed from two hydrogen atoms and one oxygen atom. This simple combination results in a substance that is essential for life, showcasing how different elements interact to create diverse materials with varying properties.

Chemical compounds can be categorized into two primary types: organic and inorganic. Organic compounds, which include carbon, are foundational to life and encompass a vast range of substances, from simple sugars to complex proteins. In contrast, inorganic compounds, which may or may not contain carbon, include minerals and metals.

Understanding chemical compounds involves not just recognizing their formulas but also comprehending how they behave in different conditions. This behavior is influenced by the nature of the bonds formed between atoms—covalent, ionic, or metallic. These bonds play a crucial role in determining a compound's reactivity, boiling and melting points, and solubility.

The Role of Organic Chemistry

Organic chemistry, a subfield of chemistry, specifically focuses on carbon-containing compounds. Carbon's unique ability to form stable bonds with a variety of elements, including itself, makes it unparalleled in its versatility. Organic chemistry is vital for several reasons: it explains the structure and behavior of biomolecules, underpins the development of pharmaceuticals, and drives innovations in materials science.

In organic chemistry, the understanding of functional groups—specific groupings of atoms that impart characteristic properties to molecules—is essential. For example, the presence of an alcohol group (-OH) can significantly alter a compound's solubility and reactivity compared to a compound without this group.

The study of organic compounds has led to groundbreaking discoveries in various fields. In medicine, organic chemistry enables the design and synthesis of drugs that can target specific biological pathways. The ability to manipulate molecular structures has profound implications for treating diseases, managing pain, and improving overall health.

As we delve deeper into the specifics of $C_{10}H_{15}N$, it is crucial to recognize its roots in organic chemistry. This compound, often associated with various applications and cultural significance, exemplifies how organic chemistry can influence both science and society.

Conclusion

Chapter 1 lays the foundation for understanding the intricate world of chemical compounds, particularly within the realm of organic chemistry. By appreciating the basic principles of chemistry, readers can better grasp the significance of $C_{10}H_{15}N$ and its impact across various domains. In the chapters that follow, we will explore the discovery of this compound, its chemical properties, and its diverse applications, ultimately revealing the remarkable interplay between chemistry and culture.

Chapter 2: The Discovery of $C_{10}H_{15}N$

Historical Context

The story of $C_{10}H_{15}N$ is one woven into the fabric of scientific discovery, cultural upheaval, and changing societal values. This compound, commonly known as methamphetamine, has a complex history that reflects humanity's ongoing quest for knowledge, innovation, and understanding of the substances that influence our lives.

The early 20th century was a period of rapid advancement in chemistry and pharmacology. In 1887, a German chemist named **Ludwig L. Ferrein** first synthesized methamphetamine from ephedrine, a compound derived from the Ephedra plant. However, it wasn't until the 1920s that methamphetamine began to attract serious attention. Pharmaceutical companies, inspired by the potential medicinal benefits of this new compound, began to explore its applications. By 1932, the compound was commercially synthesized and branded as "Pervitin" in Germany, touted as a miracle drug for a range of ailments, including asthma and depression.

The use of C10H15N rapidly expanded during World War II when it was administered to soldiers to combat fatigue and enhance performance. Military forces from multiple countries recognized its potential as a stimulant, leading to widespread use among troops. The aftermath of the war saw an increased fascination with the compound in civilian life, where it was marketed for various purposes, from weight loss to improving mood.

However, the burgeoning popularity of methamphetamine also led to emerging concerns about addiction and abuse. By the 1950s, reports of dependency and negative health effects began to surface, prompting government action. This complex interplay of discovery and misuse set the stage for the legal and cultural battles that would ensue in the decades that followed.

Key Figures in the Discovery

The discovery and subsequent popularization of C10H15N involved several key figures whose contributions significantly shaped its narrative.

1. **Ludwig L. Ferrein**: As mentioned, Ferrein was the first to synthesize methamphetamine. His work laid the groundwork for future studies and applications of this compound, although the societal implications of his discovery were not fully realized at the time.
2. **Otto Eisels**: In the early 1930s, Eisels and his colleagues were instrumental in developing Pervitin. Their research focused on the compound's effects on the human body, promoting its use as a performance enhancer and treatment for various medical conditions.
3. **Dr. David H. Houghton**: A psychiatrist who studied the psychological effects of methamphetamine in the mid-20th century, Houghton's research contributed to the understanding of the drug's potential for addiction and its mental health impacts. His warnings about the dangers of methamphetamine use foreshadowed the future challenges society would face.
4. **The FDA and Government Regulators**: As the negative repercussions of methamphetamine use became evident, regulatory bodies began to take action. The United States Food and Drug Administration (FDA) played a pivotal role in classifying methamphetamine and imposing regulations that shaped its availability and legal status.

These individuals, among many others, played crucial roles in the narrative of C10H15N, contributing to both its medical applications and the growing awareness of its potential for harm. Their efforts illustrate the dual-edged nature of scientific discovery—while it can lead to remarkable advancements, it can also create unforeseen challenges that society must navigate.

Conclusion

Chapter 2 explored the historical context and key figures involved in the discovery of C10H15N. From its synthesis in the late 19th century to its widespread use during World War II and the subsequent regulatory challenges, the narrative of this compound is a testament to the complexities of scientific innovation. As we progress to the next chapter, we will delve deeper into the chemical structure and properties of C10H15N, providing a foundational understanding of how its unique characteristics contribute to its diverse applications and cultural significance.

Chapter 3: Chemical Structure and Properties

Molecular Composition

To understand C10H15N—commonly known as methamphetamine—it is essential to explore its molecular structure. The formula C10H15N indicates that each molecule comprises ten carbon (C) atoms, fifteen hydrogen (H) atoms, and one nitrogen (N) atom. This simple yet intricate combination allows for a range of behaviors and characteristics that define the compound's interactions in both biological and chemical contexts.

Carbon Backbone

The carbon atoms in C10H15N form the backbone of the molecule, creating a structure known as a hydrocarbon. Carbon's ability to form four covalent bonds with other atoms allows for the creation of complex, three-dimensional shapes. In methamphetamine, the carbon chain is organized in a specific manner, contributing to the compound's physical and chemical properties.

The presence of nitrogen within the structure introduces basic properties, as nitrogen can form three bonds and possesses a lone pair of electrons. This not only affects how the molecule interacts with other substances but also influences its solubility and reactivity, especially in biological systems.

Physical and Chemical Properties

Understanding the physical and chemical properties of $C_{10}H_{15}N$ provides insights into its potential applications and implications for health and society.

Physical Properties

1. **Appearance**: Methamphetamine is typically a white, odorless, crystalline solid, which contributes to its illicit street form known as "crystal meth." This appearance plays a significant role in its attractiveness to users.
2. **Solubility**: Methamphetamine is soluble in water, alcohol, and other organic solvents. This solubility facilitates its absorption in the body, allowing for rapid onset of effects when consumed.
3. **Melting and Boiling Points**: Methamphetamine has a melting point of approximately 170 °C (338 °F) and a boiling point of about 215 °C (419 °F). These thermal properties are relevant for both its synthesis and its behavior during use.

Chemical Properties

1. **Stability and Reactivity**: C10H15N is relatively stable under standard conditions but can undergo oxidation reactions. Its nitrogen atom allows for the formation of additional bonds with other elements, contributing to its reactivity in various chemical environments.
2. **Acid-Base Behavior**: Methamphetamine exhibits basic properties due to the nitrogen atom, allowing it to accept protons (H^+) in aqueous solutions. This characteristic is crucial for its interaction with biological systems, as it influences the drug's effectiveness and the duration of its effects.
3. **Interactions with Biological Molecules**: The unique structure of C10H15N allows it to interact with neurotransmitter systems in the brain, particularly those involving dopamine, norepinephrine, and serotonin. This interaction is central to understanding both the therapeutic effects and the potential for abuse associated with the compound.

Conclusion

Chapter 3 has provided an overview of the molecular composition and properties of $C_{10}H_{15}N$. By dissecting its structure and exploring its physical and chemical characteristics, we gain a clearer understanding of how this compound functions in various contexts. As we move to the next chapter, we will explore the synthesis of $C_{10}H_{15}N$, examining the methods used in its creation and the laboratory techniques that highlight the complexities of working with this powerful substance. This foundation will further illuminate the balance between its medicinal potential and the challenges posed by its misuse.

Chapter 4: The Synthesis of $C_{10}H_{15}N$

Methods of Synthesis

The synthesis of $C_{10}H_{15}N$, or methamphetamine, involves various chemical processes that can yield the compound through different routes. Each method has distinct implications for both legitimate pharmaceutical use and illicit production. Understanding these methods is crucial for comprehending the complexities surrounding methamphetamine.

1. Reductive Amination of Phenylacetone

One of the most common methods for synthesizing methamphetamine is through reductive amination, where phenylacetone is reacted with ammonia and a reducing agent. This method has been favored in both research and illicit labs due to its relative efficiency.

Process Overview

- **Formation of Imine**: Phenylacetone is mixed with ammonia to form an imine.
- **Reduction**: The imine is then reduced using hydrogen gas in the presence of a metal catalyst, such as palladium or platinum, to yield methamphetamine.

This route is appreciated for its straightforwardness and high yield, making it attractive for various applications.

2. Reduction of Ephedrine/Pseudoephedrine

Another widely known method involves the reduction of ephedrine or pseudoephedrine, which are often found in over-the-counter cold medications. This process is particularly notorious for its accessibility and simplicity.

Process Overview

- **Extraction**: Ephedrine or pseudoephedrine is extracted from commercial formulations.
- **Reduction**: The compound is treated with a reducing agent like lithium aluminum hydride or red phosphorus with iodine, resulting in the conversion to methamphetamine.

This method has garnered significant attention due to its role in the rise of illicit meth production, especially in home laboratories.

3. Benzyl Methyl Ketone (BMK) Synthesis

The BMK synthesis route is another approach, which is more complex but can be employed in larger-scale operations.

Process Overview

- **Starting Material**: This method typically begins with benzyl chloride.
- **Reactions**: A series of chemical reactions involving alkylation and reduction ultimately yield methamphetamine.

While this method requires more sophisticated techniques and materials, it is often used by those producing methamphetamine on a larger scale.

Laboratory Techniques

The synthesis of methamphetamine in a controlled laboratory setting requires precise techniques and equipment to ensure both safety and efficiency. Here are key laboratory techniques involved in the synthesis process:

1. Chromatography

Chromatography is an essential technique for purifying methamphetamine and isolating it from by-products. Both gas chromatography (GC) and high-performance liquid chromatography (HPLC) are commonly used in research settings.

Purpose

2. Spectroscopy

Spectroscopic techniques, such as infrared (IR) and nuclear magnetic resonance (NMR) spectroscopy, are vital for confirming the identity of synthesized methamphetamine.

Application

3. Safety Protocols

Given the potential hazards associated with methamphetamine synthesis, adherence to safety protocols is paramount.

- **Personal Protective Equipment (PPE)**: Lab coats, gloves, goggles, and appropriate respirators are crucial to prevent exposure to toxic chemicals.
- **Ventilation**: Working in fume hoods ensures that harmful vapors are safely vented away from the laboratory environment.
- **Chemical Waste Disposal**: Proper disposal of chemical waste is necessary to minimize environmental impact and ensure compliance with regulations.

Conclusion

Chapter 4 examined the methods and laboratory techniques involved in synthesizing C10H15N. By exploring the reductive amination, reduction of ephedrine, and the BMK route, we gain insight into the processes that can yield this powerful compound. Moreover, understanding the laboratory techniques and safety protocols highlights the importance of responsible handling of such substances.

As we transition to the next chapter, we will delve into the applications of C10H15N in medicine, exploring its therapeutic uses and the challenges posed by its potential for abuse. This exploration will reveal the dual nature of this compound, as both a tool for healing and a substance of concern in public health discussions.

Chapter 5: Applications in Medicine

Medical Uses and Discoveries

$C_{10}H_{15}N$, known more commonly as methamphetamine, has a complex and multifaceted role in the field of medicine. While often associated with illicit use and addiction, methamphetamine also possesses legitimate medical applications that underscore its potential benefits when used responsibly. This chapter explores its therapeutic uses, historical context, and the impact of these applications on healthcare.

1. Therapeutic Uses

Methamphetamine is recognized in medical settings primarily for its efficacy in treating attention deficit hyperactivity disorder (ADHD) and certain cases of obesity.

- **ADHD Treatment**: Methamphetamine, under the brand name Desoxyn, is prescribed to help manage ADHD symptoms. It works by increasing the levels of neurotransmitters in the brain, particularly dopamine and norepinephrine, which play crucial roles in attention and focus. Clinical studies have demonstrated that, when administered in controlled doses, methamphetamine can significantly improve concentration, impulse control, and overall behavior in children and adults with ADHD.
- **Obesity Management**: Methamphetamine is also prescribed as part of a comprehensive treatment plan for obesity in individuals who have not successfully lost weight through diet and exercise alone. It acts as an appetite suppressant, helping patients to reduce caloric intake and promote weight loss. However, this application is tightly regulated due to the potential for misuse and addiction.

2. Case Studies of Impact

Numerous case studies illustrate the effective use of methamphetamine in medical practice, highlighting both its benefits and challenges.

- **Case Study 1: ADHD Management**: A clinical trial involving children diagnosed with ADHD showed that those treated with Desoxyn experienced significant improvements in attention span and a reduction in hyperactive behaviors. The results underscored the importance of careful monitoring and dosage adjustments, emphasizing that, while methamphetamine can be effective, it must be used judiciously.
- **Case Study 2: Weight Loss Treatment**: In a study focusing on patients with obesity, those administered methamphetamine as part of a structured weight-loss program showed notable decreases in body mass index (BMI) over six months. Participants reported improved adherence to dietary guidelines, highlighting the drug's role in supporting lifestyle changes. However, the study also raised concerns about long-term dependency, prompting discussions on patient selection and monitoring.

3. Risks and Considerations

Despite its medical applications, the use of methamphetamine is fraught with risks, necessitating stringent guidelines and monitoring.

- **Potential for Abuse**: The same properties that make methamphetamine effective for treating ADHD and obesity also contribute to its potential for abuse. The drug's stimulant effects can lead to euphoria, prompting misuse among individuals seeking recreational effects.
- **Side Effects**: Common side effects of methamphetamine include increased heart rate, anxiety, insomnia, and potential for cardiovascular complications. These risks necessitate careful patient evaluation and ongoing monitoring.
- **Regulation and Prescription**: Due to its potential for abuse, methamphetamine is classified as a Schedule II controlled substance in the United States. This classification restricts its availability and requires healthcare providers to adhere to strict prescribing guidelines.

4. Emerging Research and Future Directions

Current research is exploring new therapeutic avenues for methamphetamine, aiming to expand its benefits while minimizing risks.

- **Neuroprotective Properties**: Recent studies suggest that methamphetamine may have neuroprotective effects in certain neurological conditions, such as traumatic brain injury and neurodegenerative diseases. Preliminary findings indicate that methamphetamine could enhance recovery and cognitive function in affected individuals, paving the way for innovative treatment protocols.
- **Combination Therapies**: Research is also focusing on combining methamphetamine with other pharmacological agents to enhance its therapeutic effects while mitigating side effects. These combination therapies may provide a more balanced approach to treatment, catering to the specific needs of patients.

Conclusion

Chapter 5 has examined the medical applications of $C_{10}H_{15}N$, highlighting its legitimate uses in treating ADHD and obesity, while acknowledging the associated risks and regulatory measures. Through case studies, we see the balance between the therapeutic benefits and the potential for misuse, underscoring the importance of responsible prescribing practices. As we move into the next chapter, we will explore the cultural significance of $C_{10}H_{15}N$, investigating its impact on art, music, and literature, and how these influences shape public perception and understanding of this complex compound.

Chapter 6: Cultural Significance

$C_{10}H_{15}N$ in Pop Culture

The influence of $C_{10}H_{15}N$, more commonly known as methamphetamine, extends far beyond its scientific and medical applications. This compound has permeated various facets of culture, including literature, film, music, and visual arts, each reflecting societal attitudes and perceptions surrounding the drug. In this chapter, we explore how $C_{10}H_{15}N$ has been represented in popular culture and how these portrayals impact public understanding and discourse.

1. Film and Television

$C_{10}H_{15}N$ has found a prominent place in film and television, often depicted in narratives that explore addiction, crime, and the complexities of the human experience.

- **"Breaking Bad"**: Perhaps the most significant cultural representation of methamphetamine is in the acclaimed television series *Breaking Bad*. The show follows Walter White, a high school chemistry teacher turned methamphetamine manufacturer. Through its intense storytelling and character development, *Breaking Bad* highlights the drug's destructive potential, the moral dilemmas faced by individuals involved in its production, and the consequences for families and communities. This series has shaped public perception, humanizing the struggles of those affected by addiction while also showcasing the allure and danger of the drug trade.
- **Documentaries**: Various documentaries have also tackled the realities of methamphetamine use and its impact on society. Programs like *The Meth Epidemic* offer raw, unfiltered insights into addiction, recovery, and the socioeconomic factors contributing to the meth crisis. These portrayals serve to educate viewers about the stark realities of methamphetamine use, often countering sensationalized depictions in fictional media.

2. Literature

Literary works exploring methamphetamine often delve into the psychological and social ramifications of addiction, providing nuanced perspectives on the lives affected by the drug.

- **Fictional Narratives**: Novels like *Methland: The Death and Life of an American Small Town* by Nick Reding paint a vivid picture of the struggles faced by communities grappling with methamphetamine addiction. Reding's exploration of a small Iowa town highlights the drug's impact on families and local economies, serving as both a cautionary tale and a call for understanding and compassion.
- **Memoirs**: Personal narratives and memoirs by individuals recovering from addiction provide powerful insights into the experiences of those who have battled methamphetamine dependence. Books like *Tweak* by Nic Sheff reveal the harrowing journey through addiction, showcasing the emotional turmoil and struggles faced during recovery.

3. Music and Art

C10H15N has also found its way into music and visual arts, serving as a motif for rebellion, despair, and the complexities of addiction.

- **Musical Influence**: Songs that reference methamphetamine often encapsulate the drug's allure and destructive nature. Artists across genres, from rock to hip-hop, have addressed themes of addiction and its consequences in their lyrics. Tracks like *Methamphetamine* by the band *The Dead Kennedys* critique the societal issues surrounding drug use, while others reflect personal battles with addiction.
- **Visual Arts**: Artists have used methamphetamine as a subject to explore the interplay of addiction, identity, and societal norms. For example, contemporary art installations might incorporate materials symbolizing the drug, challenging viewers to confront uncomfortable truths about substance use and its ramifications.

4. Impact on Public Perception

The portrayal of C10H15N in popular culture significantly influences public perception and discourse surrounding methamphetamine.

- **Stigmatization and Empathy**: While some representations may contribute to the stigmatization of individuals struggling with addiction, others foster empathy and understanding. Shows like *Breaking Bad* can lead to conversations about the socioeconomic factors contributing to drug production and use, prompting viewers to reconsider their preconceived notions about addiction.
- **Awareness and Education**: Cultural representations can also serve as powerful tools for education, raising awareness about the risks and realities of methamphetamine use. By portraying the consequences of addiction, media can encourage discussions about prevention, recovery, and support systems for those affected.

Conclusion

Chapter 6 has explored the cultural significance of C10H15N, focusing on its representations in film, literature, music, and visual arts. These portrayals shape public perceptions, influence societal attitudes, and provide insights into the complexities of addiction. As we transition into the next chapter, we will delve into the psychological aspects of C10H15N, examining how it affects the brain and the nature of the psychedelic experience associated with its use. This exploration will further illuminate the intricate relationship between C10H15N and human experience, highlighting both its allure and its dangers.

Chapter 7: The Psychedelic Experience

How C10H15N Affects the Brain

C10H15N, commonly known as methamphetamine, is often associated with intense and complex psychological effects. This chapter delves into how this compound interacts with the brain, the nature of the experiences it induces, and the psychological perspectives surrounding its use. Understanding the mechanisms behind these effects is crucial for grasping both the allure and the dangers of C10H15N.

1. Mechanism of Action

Methamphetamine primarily exerts its effects by increasing the release of dopamine in the brain, a neurotransmitter associated with pleasure, reward, and motivation. The drug achieves this through several mechanisms:

- **Dopamine Release**: Methamphetamine prompts the release of large amounts of dopamine in the brain's reward pathway. This surge creates intense feelings of euphoria and increased energy, contributing to the drug's addictive potential.
- **Inhibition of Reuptake**: By blocking the reuptake of dopamine, methamphetamine prolongs its action in the synaptic cleft. This not only enhances the euphoric effects but also disrupts normal neurotransmitter balance, leading to potential long-term changes in brain function.
- **Impact on Other Neurotransmitters**: In addition to dopamine, methamphetamine affects other neurotransmitters, including norepinephrine and serotonin. This broad impact can lead to a range of emotional and physiological responses, from heightened alertness to anxiety and agitation.

2. Psychedelic Effects and Experiences

While methamphetamine is not traditionally classified as a psychedelic, its intense effects on perception and mood can resemble those of psychedelics in some respects. Users often report a range of experiences that can include:

- **Euphoria and Energy**: Many users initially experience a powerful sense of euphoria, increased sociability, and heightened energy levels. This "rush" is often sought after and can lead to repeated use.
- **Altered Perception**: Some users describe altered sensory perceptions, including enhanced visual and auditory stimuli. This alteration can create a feeling of connectedness or heightened awareness, resembling experiences reported by users of classic psychedelics.
- **Distortions of Reality**: Extended use can lead to significant distortions in reality, including paranoia, hallucinations, and delusions. These effects can be distressing and may lead to dangerous behaviors.

3. Psychological Perspectives

The psychological effects of $C_{10}H_{15}N$ have prompted various perspectives within the field of psychology and addiction studies. Understanding these perspectives is essential for addressing the complexities of methamphetamine use:

- **Addiction as a Brain Disease**: Many researchers view methamphetamine addiction through the lens of neurobiology, emphasizing the drug's impact on brain structure and function. Chronic use can lead to lasting changes in brain circuits associated with reward, motivation, and decision-making, complicating recovery efforts.
- **Behavioral and Cognitive Impacts**: Psychologists also focus on the behavioral and cognitive effects of methamphetamine use. Users may experience impaired judgment, increased impulsivity, and difficulty in regulating emotions, all of which can contribute to the cycle of addiction.
- **Coping Mechanisms**: Some individuals may turn to methamphetamine as a means of coping with underlying psychological issues, such as trauma or depression. Understanding these underlying factors is crucial for developing effective treatment strategies.

4. Long-Term Effects on Mental Health

The long-term use of $C_{10}H_{15}N$ can lead to severe mental health issues, further complicating the recovery process:

- **Cognitive Decline**: Research indicates that chronic methamphetamine use can result in cognitive deficits, including impairments in memory, attention, and executive function. These cognitive challenges can persist long after cessation of use.
- **Mental Health Disorders**: Users are at increased risk for various mental health disorders, including anxiety, depression, and psychosis. The relationship between methamphetamine use and mental health is complex and bidirectional, with each factor influencing the other.
- **Recovery and Rehabilitation**: Addressing the psychological effects of methamphetamine is essential for successful recovery. Effective treatment often involves a combination of behavioral therapies, support groups, and, in some cases, pharmacological interventions aimed at alleviating withdrawal symptoms and reducing cravings.

Conclusion

In this chapter, we have explored how C10H15N affects the brain and the nature of the experiences associated with its use. The intense euphoria and altered perceptions may draw individuals to the drug, but the psychological consequences can be severe and long-lasting. As we move into the next chapter, we will examine the legal and ethical considerations surrounding C10H15N, including the societal impacts of its regulation and use. Understanding these aspects is crucial for fostering informed discussions about this complex compound and its role in contemporary society.

Chapter 8: Legal and Ethical Considerations

Legislation Surrounding C10H15N

The legal landscape surrounding C10H15N (methamphetamine) is complex and reflects the substance's dual nature as both a potentially beneficial compound in controlled medical contexts and a dangerous drug when misused. Understanding this legal framework is crucial for grasping the societal implications of C10H15N, its regulation, and the ongoing debates about its use.

1. Historical Context of Legislation

Methamphetamine was first synthesized in 1893, but its widespread misuse began during World War II, when it was used to keep soldiers alert. After the war, concerns about its addictive potential led to increased scrutiny and regulation.

- **Controlled Substances Act**: In the United States, the 1970 Controlled Substances Act classified methamphetamine as a Schedule II drug, recognizing its medical uses but also its high potential for abuse. This classification places stringent controls on its prescription and distribution, limiting its availability outside of medical contexts.
- **State Legislation**: Various states have enacted laws to combat the methamphetamine epidemic, including restrictions on pseudoephedrine (a precursor in methamphetamine synthesis) and increased penalties for trafficking. These laws aim to reduce access while addressing the public health crisis associated with methamphetamine addiction.

2. Current Legal Framework

Today, methamphetamine is primarily prescribed in low doses for conditions such as ADHD and obesity, under strict medical supervision. However, its illicit use remains a significant concern, leading to ongoing legislative efforts:

- **Prescription Regulations**: Medical professionals prescribing methamphetamine must adhere to strict guidelines, ensuring that patients are closely monitored for signs of misuse or addiction.
- **Criminal Penalties**: The manufacture, distribution, and possession of methamphetamine without a prescription carry severe legal consequences, reflecting its status as a controlled substance. Offenders may face imprisonment, heavy fines, and mandatory rehabilitation programs.

3. Global Perspectives on Legislation

The legal status of $C_{10}H_{15}N$ varies globally, reflecting differing cultural attitudes and public health strategies:

- **International Treaties**: Various international treaties, such as the United Nations Convention against Illicit Traffic in Narcotic Drugs and Psychotropic Substances, aim to regulate the production and distribution of controlled substances, including methamphetamine. These treaties necessitate cooperation among nations to combat drug trafficking and promote public health.
- **Diverse Regulations**: Countries like Japan and South Korea have particularly stringent laws against methamphetamine, shaped by historical abuse and social stigma. In contrast, some nations are exploring more lenient approaches, focusing on harm reduction and rehabilitation rather than punitive measures.

Ethical Debates in Use

The conversation surrounding $C_{10}H_{15}N$ is not solely about legality; it also encompasses profound ethical questions regarding its medical use, potential benefits, and the responsibilities of healthcare providers and society at large.

1. Medical vs. Recreational Use

- **Access to Treatment**: One ethical consideration is the balance between ensuring access to methamphetamine for legitimate medical uses and preventing its misuse. Healthcare providers must navigate the fine line between offering effective treatment for conditions like ADHD and minimizing the risk of addiction.
- **Informed Consent**: Patients receiving methamphetamine prescriptions must be fully informed about the potential risks and benefits. Ethical practice requires that patients understand their treatment options and the implications of using a controlled substance.

2. Public Health vs. Criminal Justice

- **Harm Reduction Strategies**: Some advocates argue for a shift from criminalization to harm reduction approaches, emphasizing treatment over punishment for individuals struggling with methamphetamine addiction. This perspective encourages the development of comprehensive public health strategies that address addiction as a health issue rather than a criminal one.
- **Resource Allocation**: Ethical debates also arise around resource allocation for addiction treatment versus law enforcement. Decisions about funding and support can reflect societal values regarding addiction, public health, and the stigmatization of drug use.

3. Social Implications

- **Stigmatization**: The stigma associated with methamphetamine use can have profound social consequences. Individuals who struggle with addiction may face discrimination, making it more challenging to seek help and reintegrate into society.
- **Community Impact**: The effects of methamphetamine addiction ripple through communities, impacting families, social services, and local economies. Ethical considerations must account for the broader societal implications of both addiction and legislative responses.

Conclusion

The legal and ethical landscape surrounding C10H15N is multifaceted, reflecting the complexities of its medical use and the societal challenges posed by its illicit consumption. As we move into the next chapter, we will explore the role of modern research in advancing our understanding of C10H15N, its therapeutic potential, and the implications for future legislation and public health strategies. Understanding the interplay of law, ethics, and science is crucial for addressing the ongoing challenges associated with this powerful compound.

Chapter 9: The Role in Modern Research

Current Studies and Innovations

As research into C10H15N (methamphetamine) evolves, it increasingly illuminates the compound's complexities and potential applications beyond its notorious reputation. Modern studies are exploring its therapeutic benefits, mechanisms of action, and the biological pathways involved, with a focus on responsible use in medical contexts.

1. Therapeutic Research

Recent investigations are shifting towards understanding how C10H15N can be utilized in treating specific medical conditions:

- **Attention Deficit Hyperactivity Disorder (ADHD)**: Clinical studies have established the effectiveness of methamphetamine as a treatment for ADHD. Researchers are examining dosage, long-term effects, and comparative efficacy against other stimulant medications, aiming to refine treatment protocols and improve patient outcomes.
- **Obesity Management**: Methamphetamine's appetite-suppressing properties are being revisited as potential tools in combating obesity. Research is focusing on its long-term effectiveness and safety in weight management strategies, as well as how it compares with other pharmacological options.

2. Mechanisms of Action

Understanding how $C_{10}H_{15}N$ interacts with the brain is a focal point of current research:

- **Dopamine Pathways**: Methamphetamine primarily affects the central nervous system by increasing the release of dopamine. Studies are delving into the neurobiological mechanisms that govern this process, exploring how it can be harnessed for therapeutic benefits while minimizing the risk of addiction.
- **Neuroprotection and Neuroplasticity**: Some studies suggest that low doses of methamphetamine may have neuroprotective effects. Researchers are investigating its potential role in enhancing neuroplasticity, which could have implications for recovery from neurological disorders.

3. Innovative Delivery Methods

The route of administration significantly impacts the pharmacokinetics and safety profile of C10H15N. Recent research is focusing on innovative delivery methods:

- **Extended-Release Formulations**: Developing formulations that allow for extended-release of the compound could reduce the potential for misuse and side effects. Research is examining how these methods can stabilize plasma levels, providing therapeutic benefits while minimizing peaks and troughs associated with traditional dosing.
- **Transdermal Patches and Inhalers**: Alternative delivery systems, such as transdermal patches or inhalers, are being studied for their potential to provide controlled and consistent dosing. This approach could reduce the stigma associated with oral consumption and provide a more discreet option for patients.

Future Directions in Research

The future of research involving C10H15N is poised to address several critical areas, focusing on both therapeutic potential and public health implications.

1. Longitudinal Studies on Long-term Use

As the therapeutic landscape shifts, there is a pressing need for longitudinal studies that examine the long-term effects of prescribed methamphetamine use. Such studies could provide vital data on:

- **Adverse Effects**: Understanding the potential long-term impacts on mental health, cardiovascular health, and overall well-being is crucial for informing treatment guidelines.
- **Quality of Life Measures**: Evaluating how treatment with $C_{10}H_{15}N$ influences patients' daily lives, relationships, and productivity will provide a holistic view of its benefits and risks.

2. Interdisciplinary Approaches

Future research will likely benefit from interdisciplinary collaboration, integrating insights from various fields:

- **Psychiatry and Neuroscience**: Collaborative studies between psychiatrists and neuroscientists could enhance understanding of the compound's psychological effects and its role in the brain's reward pathways.
- **Public Health and Policy**: Engaging public health experts in research will be vital to ensure that findings inform policies that balance therapeutic benefits with the need for regulation and public safety.

3. Addressing Stigma and Public Perception

Research efforts must also focus on addressing the stigma associated with $C_{10}H_{15}N$:

- **Community Engagement**: Initiatives aimed at educating the public about the potential medical uses of methamphetamine can help shift perceptions, fostering a more informed discussion around its benefits and risks.
- **Advocacy for Responsible Use**: Promoting responsible use through educational campaigns can support efforts to integrate $C_{10}H_{15}N$ into accepted medical practice while emphasizing the importance of regulation and oversight.

Conclusion

The role of $C_{10}H_{15}N$ in modern research highlights a transformative shift in understanding this compound. As studies progress, they are uncovering potential therapeutic applications and informing public health strategies. The interplay between innovation and responsibility will be key as we navigate the future of $C_{10}H_{15}N$, balancing its medical potential with societal implications. In the next chapter, we will explore the societal impacts of $C_{10}H_{15}N$, examining how various communities perceive and are influenced by this complex compound.

Chapter 10: Societal Impacts

The compound C10H15N, known more commonly as methamphetamine, has a multifaceted influence on society that extends beyond its medical applications and cultural portrayals. As communities grapple with its presence and implications, understanding the societal impacts of this compound is crucial for fostering informed discussions and responsible practices. This chapter explores how C10H15N affects various communities, the perspectives of different groups, and the broader social implications tied to its use.

Influence on Communities

1. Health Impacts

One of the most significant societal impacts of C10H15N is on public health. Communities affected by methamphetamine use often face a range of health issues, from addiction and mental health disorders to increased rates of communicable diseases due to unsafe practices.

- **Addiction and Recovery**: Many communities are experiencing the burden of addiction, leading to increased demand for treatment services. Organizations focused on recovery have emerged, offering support and resources for those struggling with addiction. These services are critical in mitigating the negative health impacts on families and neighborhoods.
- **Mental Health**: The psychological effects of methamphetamine use, including anxiety, paranoia, and depression, have significant repercussions for community mental health. Increased awareness and resources for mental health support are essential in addressing these challenges.

2. Economic Consequences

The economic implications of $C_{10}H_{15}N$ are profound, affecting local economies, law enforcement resources, and healthcare systems.

- **Law Enforcement Costs**: Communities face significant expenses related to drug enforcement and crime prevention. The presence of methamphetamine often correlates with higher crime rates, requiring more extensive policing and community safety initiatives.
- **Healthcare System Strain**: The healthcare system in areas with high rates of methamphetamine use can become overwhelmed. Emergency services, addiction treatment facilities, and mental health resources all bear the burden, leading to increased healthcare costs and reduced access to care for other conditions.

3. Social Stigma

The stigma surrounding $C_{10}H_{15}N$ use can perpetuate cycles of marginalization and hinder recovery efforts.

- **Public Perception**: Individuals struggling with methamphetamine addiction often face societal judgment, leading to feelings of shame and isolation. This stigma can discourage people from seeking help and participating in community recovery programs.
- **Community Divisions**: Stigmatization can create divisions within communities, as residents may fear association with drug-related activities. Fostering understanding and compassion is essential for promoting inclusive community responses to substance use.

Perspectives from Various Groups

1. Public Health Advocates

Public health advocates emphasize the need for education, prevention, and treatment strategies focused on $C_{10}H_{15}N$. They advocate for:

- **Comprehensive Education Programs**: Initiatives aimed at raising awareness about the risks and realities of methamphetamine use are crucial. These programs often target schools, community centers, and healthcare providers to disseminate information effectively.
- **Harm Reduction Strategies**: Advocates support harm reduction approaches, such as needle exchange programs and safe consumption sites, to mitigate the negative impacts of methamphetamine use. These strategies can help reduce the spread of infectious diseases and provide pathways to recovery.

2. Community Organizations

Local organizations play a vital role in addressing the challenges posed by $C_{10}H_{15}N$. Their efforts often include:

- **Support Groups**: Community-led support groups offer a safe space for individuals and families affected by methamphetamine use. These groups provide resources, share experiences, and foster a sense of belonging.
- **Employment and Rehabilitation Programs**: Organizations focused on rehabilitation often incorporate job training and placement services, helping individuals reintegrate into society and reduce the likelihood of relapse.

3. Law Enforcement

Law enforcement perspectives on C10H15N often emphasize the need for a balanced approach:

- **Focus on Prevention**: Many law enforcement agencies are shifting their focus from punitive measures to prevention and treatment options. This includes collaborating with public health initiatives and community organizations to address the root causes of drug use.
- **Community Policing**: Emphasizing community-oriented policing strategies can help build trust between law enforcement and residents, fostering cooperation in tackling methamphetamine-related issues.

Conclusion

The societal impacts of C10H15N are complex, intertwining public health, economic, and social factors. As communities navigate the challenges posed by this compound, collaboration among public health advocates, community organizations, law enforcement, and individuals is essential.

By fostering understanding and promoting compassionate responses, society can work towards effective solutions that address the realities of C10H15N while supporting those affected by its use. In the following chapter, we will explore the environmental considerations related to C10H15N, examining its ecological footprint and sustainable practices associated with its production and use.

Chapter 11: C10H15N and the Environment

The environmental implications of C10H15N, commonly associated with methamphetamine, extend beyond human health concerns and touch on ecological impacts as well. Understanding these environmental factors is crucial for developing sustainable practices and mitigating harm. This chapter explores the ecological footprint of C10H15N, the environmental consequences of its production and use, and sustainable practices that can help reduce its negative impact.

Ecological Footprint

1. Production and Manufacturing

The synthesis of C10H15N, particularly in illicit contexts, often involves hazardous chemicals and processes that can have dire environmental consequences.

- **Toxic Waste**: Illegal methamphetamine labs frequently produce toxic waste, including solvents, acids, and other chemical byproducts. Improper disposal of these substances can contaminate soil and water sources, posing risks to local ecosystems and communities.
- **Land and Water Contamination**: The residue from meth production can leach into the ground, leading to long-term soil contamination. Aquatic ecosystems can be affected when hazardous materials enter waterways, impacting flora and fauna and potentially entering the food chain.

2. Land Use Changes

The production of C10H15N often leads to land use changes, particularly in rural areas where illicit labs may be established.

- **Deforestation and Habitat Loss**: In some regions, the establishment of meth labs may coincide with deforestation or land clearing, disrupting local habitats and biodiversity. The removal of trees and vegetation can also contribute to soil erosion and altered water cycles.
- **Increased Crime and Instability**: The presence of drug manufacturing can destabilize communities, leading to increased crime and social unrest, which may further exacerbate environmental degradation.

Sustainable Practices

1. Reducing Chemical Use

Promoting safer and more sustainable chemical practices can significantly mitigate the environmental impact associated with $C_{10}H_{15}N$.

- **Green Chemistry**: Adopting green chemistry principles in research and industry can minimize hazardous substances and reduce waste. By designing processes that utilize less harmful chemicals and generate less waste, the overall environmental footprint can be reduced.
- **Regulation of Precursor Chemicals**: Effective regulation of precursor chemicals used in the synthesis of $C_{10}H_{15}N$ can help curb illicit production. Implementing stricter controls and promoting alternative methods can reduce the availability of these substances for illegal use.

2. Cleanup and Remediation

Addressing the legacy of past methamphetamine production requires concerted cleanup efforts.

- **Environmental Remediation**: Initiatives focused on environmental remediation can help restore contaminated sites. This may involve soil testing, removal of hazardous materials, and replanting native vegetation to restore habitats.
- **Community Involvement**: Engaging communities in cleanup efforts fosters a sense of ownership and responsibility. Community-led initiatives can mobilize resources and raise awareness about the environmental impacts of $C_{10}H_{15}N$.

3. Education and Awareness

Raising awareness about the environmental consequences of $C_{10}H_{15}N$ is essential for promoting sustainable practices.

- **Public Education Campaigns**: Implementing educational campaigns can inform the public about the ecological risks associated with methamphetamine production and use. This awareness can encourage more responsible behaviors and support for environmental protection.
- **Integration into Curriculum**: Incorporating environmental education into school curricula can empower the next generation with knowledge about chemistry and sustainability, equipping them to make informed decisions.

Conclusion

The environmental implications of C10H15N underscore the need for a multifaceted approach to address its impact. By understanding the ecological footprint of its production and use, communities can implement sustainable practices that not only mitigate harm but also promote recovery and restoration.

As we transition to the next chapter, we will delve into personal accounts and testimonials, sharing diverse perspectives from users, researchers, and community members. These narratives will highlight the human aspect of the C10H15N experience, providing insights into its impact on individual lives and communities.

Chapter 12: Personal Accounts and Testimonials

The impact of C10H15N—commonly known as methamphetamine—extends far beyond its chemical structure or clinical applications. This chapter presents personal accounts and testimonials from individuals affected by this compound, including users, researchers, and community advocates. Their stories provide a deeper understanding of the multifaceted nature of C10H15N, illustrating both its potential benefits and the challenges associated with its use.

Voices of Experience

1. User Testimonials

Many individuals who have experimented with or been dependent on C10H15N share stories that reveal the compound's allure and the profound consequences of its use.

Sara, 28, Former User:

"I was introduced to meth in college. At first, it felt like a miracle. I could study all night, focus like never before, and feel invincible. But it wasn't long before I realized it was consuming me. My relationships fell apart, and I lost my job. It's a rollercoaster—an exhilarating high that inevitably leads to a devastating crash."

Tom, 35, Recovery Advocate:

"After years of battling addiction, I found help through a recovery program. What struck me was how much C10H15N affected not just me, but everyone around me. My family suffered, and it took years to rebuild those bonds. Now, I work to help others understand that there is hope, even when it feels like there isn't."

2. Researchers' Perspectives

Researchers and professionals in the field of addiction and mental health provide insights into the complexities of C10H15N's effects on individuals and communities.

Dr. Emily Chen, Psychologist:

"From a clinical perspective, C10H15N is fascinating yet troubling. It alters brain chemistry, often exacerbating underlying mental health issues. We're seeing a rise in cases where individuals struggle with both addiction and mental health disorders, creating a cycle that is difficult to break."

Dr. Raj Patel, Chemist and Researcher:

"In my work, I've focused on understanding the pharmacological effects of C10H15N. While it has potential medicinal applications, the risks of misuse and dependency are significant. It's essential to balance the research on its benefits with a thorough understanding of the societal implications."

3. Community Impact

Community advocates share how $C_{10}H_{15}N$ has influenced local environments, highlighting efforts to combat its negative effects and promote recovery.

Maria, Community Organizer:

"In our town, the rise of meth use has changed everything. We've seen families torn apart and neighborhoods destabilized. But we're not giving up. We've launched awareness campaigns, partnered with local health services, and created safe spaces for recovery support. It's a long road, but together we're making progress."

James, Law Enforcement Officer:

"As someone who has worked on the front lines of drug enforcement, I've seen firsthand the devastation that meth can cause. However, I've also witnessed incredible resilience. Communities can come together to educate, support, and create change. It's not just about policing; it's about healing and prevention."

Themes Emerging from Testimonials

1. Dual Nature of $C_{10}H_{15}N$

The accounts reveal a dual nature of C10H15N: its potential for initial euphoria and productivity contrasted sharply with the destructive consequences of addiction. Many users describe a seductive initial experience that rapidly devolves into dependency and chaos.

2. Importance of Support Systems

Many stories emphasize the critical role of support systems in recovery. Whether through family, friends, or community organizations, having a network can significantly influence outcomes. Recovery is often framed as a collective journey rather than an individual struggle.

3. Need for Education and Awareness

Both users and professionals highlight the importance of education and awareness in combating the stigma surrounding addiction. Understanding the science behind C10H15N can foster empathy and support for those affected, shifting the conversation from judgment to understanding.

Conclusion

Personal accounts surrounding C10H15N illustrate the complex interplay between this compound and the lives it touches. From stories of struggle and recovery to insights from researchers and advocates, these narratives highlight the urgent need for comprehensive education, effective treatment options, and community support. As we move to the next chapter, we will delve into the science of addiction, exploring the underlying mechanisms of dependency and innovative treatment approaches.

Chapter 13: The Science of Addiction

Understanding the dynamics of addiction, particularly concerning C10H15N, requires a deep dive into the biological, psychological, and social factors that contribute to dependency. This chapter explores the mechanisms of addiction, the implications for treatment, and the evolving landscape of recovery.

Understanding Dependency

1. Neurobiology of Addiction

At the core of addiction is the brain's reward system, a complex network of structures that influences motivation and pleasure. C10H15N affects neurotransmitter systems, primarily increasing levels of dopamine, which is associated with feelings of euphoria and reward.

- **Dopaminergic Pathways**: When C10H15N is consumed, it stimulates the release of dopamine in areas such as the nucleus accumbens and the prefrontal cortex. This activation not only produces immediate feelings of pleasure but also reinforces behaviors associated with drug use, creating a cycle of addiction.
- **Neuroadaptation**: With repeated use, the brain undergoes neuroadaptive changes, resulting in reduced sensitivity to dopamine. Users may find it increasingly difficult to experience pleasure from everyday activities, leading to a greater reliance on the substance to achieve any sense of reward.

2. Psychological Factors

The psychological aspects of addiction are multifaceted. Various factors can predispose individuals to develop a dependency on C10H15N:

- **Mental Health Conditions**: Individuals with underlying mental health issues, such as depression or anxiety, may turn to C10H15N as a form of self-medication. This can create a vicious cycle where the substance exacerbates the very conditions they sought to alleviate.
- **Coping Mechanisms**: For some, the use of C10H15N becomes a maladaptive coping mechanism in response to stress, trauma, or emotional pain. Understanding these underlying triggers is crucial for effective treatment.

3. Social and Environmental Influences

Addiction is not solely an individual issue; social and environmental factors play a significant role in its development and perpetuation.

- **Peer Influence**: Social circles that normalize or encourage substance use can significantly increase the risk of addiction. Peer pressure and the desire for acceptance can lead individuals to experiment with C10H15N, potentially resulting in dependency.
- **Socioeconomic Factors**: Communities with limited access to healthcare, education, and economic opportunities may experience higher rates of addiction. Economic stressors can drive individuals toward substances as a means of escape.

Treatment Approaches

1. Integrated Treatment Models

Given the complexity of addiction, treatment approaches are most effective when they integrate various modalities:

- **Behavioral Therapy**: Cognitive Behavioral Therapy (CBT) and other therapeutic approaches focus on modifying harmful thought patterns and behaviors. These therapies can help individuals develop healthier coping mechanisms and strategies to resist cravings.
- **Medication-Assisted Treatment (MAT)**: In some cases, medications can be used to manage withdrawal symptoms and reduce cravings. While specific MAT options for $C_{10}H_{15}N$ may still be under research, the use of medications that target neurotransmitter systems offers promise for treatment.

2. Support Systems and Community Resources

The role of community and support systems is vital in the recovery process:

- **Support Groups**: Programs such as Narcotics Anonymous (NA) provide individuals with peer support, shared experiences, and accountability. These groups foster a sense of community and belonging, which can be crucial for sustained recovery.
- **Rehabilitation Programs**: Comprehensive rehabilitation programs often combine medical care, psychological support, and social services. These programs address the multifaceted nature of addiction and assist individuals in rebuilding their lives.

3. Emerging Trends in Treatment

The field of addiction treatment is continually evolving, with innovative approaches gaining traction:

- **Telehealth Services**: The rise of telehealth has made treatment more accessible, allowing individuals to receive support from professionals without geographical limitations. This is particularly beneficial for those in remote areas.
- **Psychedelic-Assisted Therapy**: Emerging research into the therapeutic use of psychedelics for treating addiction suggests that these substances may help rewire the brain's reward system and provide profound insights during therapy.

Conclusion

Understanding the science of addiction related to C10H15N involves unraveling a complex interplay of neurobiology, psychology, and social influences. Effective treatment requires a holistic approach that considers all these factors, emphasizing the importance of community support and innovative strategies in recovery. As we proceed to the next chapter, we will explore global perspectives on C10H15N, examining how different cultures approach the use and regulation of this powerful compound.

Chapter 14: Global Perspectives

The use and acceptance of C10H15N vary significantly around the world, influenced by cultural attitudes, legal frameworks, and historical contexts. This chapter delves into how different countries perceive and regulate this compound, highlighting the diverse experiences and practices surrounding its use.

International Use and Acceptance

1. Cultural Attitudes Toward C10H15N

In many cultures, substances similar to C10H15N have a long history of use, often tied to spiritual practices, traditional medicine, or recreational activities. These cultural frameworks shape how societies view the substance today.

- **Indigenous Practices**: In several indigenous cultures, psychedelic substances are used in sacred rituals and healing practices. For example, certain groups in South America incorporate psychedelics into shamanic traditions, viewing them as tools for spiritual enlightenment and communal bonding.
- **Western Recreational Use**: In the West, C10H15N and similar compounds have been popularized within counterculture movements, particularly during the 1960s. This period saw a surge in interest as part of a broader quest for expanded consciousness and personal freedom.

2. Regulatory Approaches

The legal status of C10H15N varies widely across the globe, reflecting each nation's regulatory philosophy and approach to drug use.

- **Strict Prohibition**: In many countries, including the United States, C10H15N is classified as a Schedule I substance, meaning it is deemed to have no accepted medical use and a high potential for abuse. This classification limits research opportunities and shapes public perception, often framing the compound predominantly in negative terms.
- **Decriminalization and Regulation**: In contrast, some regions are moving toward decriminalization or regulated use. For example, certain cities in the U.S., like Denver and Oakland, have decriminalized the use of psilocybin mushrooms, and discussions around regulating C10H15N are gaining traction in various jurisdictions. These changes reflect a growing recognition of the potential therapeutic benefits and a shift toward harm reduction.

3. Medical and Scientific Research

The landscape of research regarding C10H15N varies globally, shaped by both regulatory frameworks and cultural openness to exploring its effects.

- **Emerging Research in Europe**: Countries like the Netherlands and Switzerland have taken progressive stances on the research and use of psychedelics. These nations host clinical trials exploring the therapeutic potential of compounds similar to $C_{10}H_{15}N$, particularly in treating mental health conditions such as PTSD and depression.
- **Innovative Studies in South America**: Researchers in South America are also investigating traditional uses of psychedelics within indigenous communities, aiming to document and understand their efficacy and cultural significance. These studies can bridge traditional knowledge and modern scientific inquiry.

Cultural Variations

1. Personal Narratives and Testimonials

Personal experiences with $C_{10}H_{15}N$ vary widely based on cultural context.

- **In the United States**: Users often share stories of transformative experiences, emphasizing insights gained during psychedelic trips. However, these narratives can be accompanied by cautionary tales about addiction and misuse, reflecting the tension between recreational use and the risks associated with unregulated consumption.
- **In Indigenous Cultures**: Testimonials from indigenous practitioners highlight the importance of ritual and community in the use of psychedelics. These narratives often focus on healing, guidance from ancestral spirits, and the interconnectedness of life, framing the use of $C_{10}H_{15}N$ as a sacred act rather than a recreational one.

2. Artistic Representations

Art and literature often reflect societal attitudes toward $C_{10}H_{15}N$, capturing both its allure and potential dangers.

- **Literature and Film**: Works that explore psychedelic experiences often oscillate between glorification and caution. Books and films can serve as powerful tools to shape public perceptions, depicting both the mystical aspects of C10H15N and the darker sides of addiction and mental health struggles.
- **Visual Arts**: Many contemporary artists use C10H15N as a theme, exploring its effects on perception and reality. These artistic expressions can challenge societal norms and encourage dialogue about substance use, mental health, and spirituality.

Conclusion

The global perspectives on C10H15N reveal a rich tapestry of cultural beliefs, regulatory frameworks, and individual experiences. As societies continue to navigate the complexities of this compound, understanding these varied contexts becomes essential for informed discussions about its potential and risks. In the next chapter, we will look toward the future of C10H15N, exploring emerging trends and potential breakthroughs in research and application.

Chapter 15: The Future of C10H15N

As we look ahead, the future of C10H15N promises to be both intriguing and complex. With ongoing research, shifting societal attitudes, and advancements in technology, the potential applications and implications of this compound continue to evolve. This chapter explores emerging trends, potential breakthroughs, and the broader implications of C10H15N in various fields.

Emerging Trends

1. Increasing Acceptance and Decriminalization

A notable trend in many regions is the gradual shift towards the acceptance and decriminalization of substances like C10H15N. This movement reflects a growing recognition of the compound's therapeutic potential and the limitations of punitive approaches to drug use.

- **Policy Changes**: Several U.S. cities and states are beginning to pass legislation aimed at decriminalizing or regulating the use of psychedelics, including $C_{10}H_{15}N$. This change is driven by mounting evidence from clinical studies that highlight the benefits of psychedelics in treating mental health disorders, such as depression and PTSD.
- **Public Discourse**: Increased media coverage and public discourse around the benefits of psychedelics are contributing to a cultural shift. Influential figures in science, medicine, and mental health are advocating for more research and open dialogue, which is gradually normalizing discussions about these substances.

2. Research Innovations and Clinical Trials

As scientific understanding of $C_{10}H_{15}N$ deepens, the focus on innovative research methodologies is becoming more pronounced.

- **Advanced Clinical Trials**: Pharmaceutical companies and research institutions are initiating rigorous clinical trials to explore the therapeutic applications of C10H15N. Studies are focusing on its efficacy in treating a range of conditions, including anxiety, addiction, and chronic pain.
- **Integrative Approaches**: Emerging research is beginning to integrate traditional practices with modern science. Studies are exploring how culturally-informed approaches can enhance therapeutic outcomes when using C10H15N in clinical settings.

3. Psychedelic-Assisted Therapy

The concept of psychedelic-assisted therapy is gaining traction in the mental health field.

- **Therapeutic Frameworks**: Practitioners are developing structured therapeutic frameworks that incorporate C10H15N as a tool for exploration and healing. This approach emphasizes the importance of set and setting, ensuring that individuals are supported in a safe and therapeutic environment.
- **Training Programs**: As interest in psychedelic therapy grows, training programs for therapists are being established. These programs aim to equip professionals with the knowledge and skills necessary to facilitate sessions safely and effectively.

Potential Breakthroughs

1. Personalized Medicine

The future of $C_{10}H_{15}N$ may also lie in the realm of personalized medicine.

- **Tailored Treatments**: Research is increasingly focusing on how individual differences—such as genetics, mental health history, and personal circumstances—affect responses to $C_{10}H_{15}N$. Understanding these factors could lead to more effective, tailored therapeutic approaches.
- **Biomarker Development**: Scientists are exploring the identification of biomarkers that predict individual responses to $C_{10}H_{15}N$, paving the way for personalized treatment plans that maximize benefits and minimize risks.

2. Neuroscience and Mechanisms of Action

Continued investigation into the neuroscience of $C_{10}H_{15}N$ is expected to yield significant insights.

- **Brain Imaging Studies**: Advanced neuroimaging techniques are being used to observe how C10H15N affects brain activity and connectivity. Understanding these mechanisms could clarify how the compound induces its effects, informing both clinical practices and safety guidelines.
- **Molecular Research**: Research into the molecular pathways activated by C10H15N may reveal new targets for drug development, potentially leading to novel therapies that mimic the beneficial effects without the psychedelic experience.

Broader Implications

1. Impact on Mental Health Paradigms

The potential integration of C10H15N into mental health treatment paradigms could shift how we approach mental illness.

- **Revolutionizing Treatment**: If successful, psychedelic-assisted therapies could revolutionize how mental health disorders are treated, moving away from traditional pharmacological approaches toward more holistic, integrative models that prioritize the individual's experience.
- **Reducing Stigma**: Greater acceptance and understanding of C10H15N may help reduce the stigma surrounding mental health treatment and substance use, encouraging more individuals to seek help and explore alternative therapies.

2. Ethical Considerations

As interest in C10H15N expands, ethical considerations will become increasingly important.

- **Access and Equity**: Discussions around access to psychedelic treatments must prioritize equity, ensuring that marginalized communities are not left behind in the therapeutic advancements.
- **Informed Consent**: Ethical frameworks for psychedelic-assisted therapies must ensure that informed consent is prioritized, emphasizing the importance of understanding potential risks and benefits.

Conclusion

The future of C10H15N is filled with promise and possibility. As research expands and societal attitudes shift, this compound could play a pivotal role in addressing some of the most pressing mental health challenges of our time. However, as we navigate this evolving landscape, it is essential to approach C10H15N with both optimism and caution, prioritizing safety, ethics, and a commitment to understanding the complexities of human experience. In the next chapter, we will explore common misconceptions about C10H15N and clarify the scientific truths surrounding it.

Chapter 16: Debunking Myths and Misconceptions

As with any compound that has garnered significant attention, $C_{10}H_{15}N$ is surrounded by a myriad of myths and misconceptions. These misunderstandings can cloud public perception, hinder scientific research, and affect policy decisions. This chapter aims to clarify common myths about $C_{10}H_{15}N$, offering scientific explanations to provide a more nuanced understanding of this complex compound.

Common Misunderstandings

1. $C_{10}H_{15}N$ is Just a Drug for Recreation

One of the most pervasive myths is that $C_{10}H_{15}N$ is merely a recreational drug with no legitimate applications. While it has been used recreationally, this oversimplification ignores the growing body of evidence supporting its therapeutic potential.

Scientific Insight

2. $C_{10}H_{15}N$ is Highly Addictive

Another misconception is that $C_{10}H_{15}N$ is as addictive as other substances, leading to the belief that its use will inevitably result in dependency.

Clarification

3. Using C10H15N Will Lead to Permanent Psychological Damage

The fear of lasting psychological harm is a significant deterrent for many considering the use of C10H15N, but this fear is often exaggerated.

Research Findings

4. C10H15N is a Cure-All

Some proponents of C10H15N might suggest that it can solve a wide array of mental health issues, leading to unrealistic expectations about its capabilities.

Nuanced Reality

Scientific Clarifications

1. Mechanisms of Action

A common misunderstanding is the lack of clarity around how C10H15N affects the brain.

Neuroscientific Insights

2. Cultural Context and Historical Use

Many people may not realize that C10H15N has a rich history of use in various cultures, often for spiritual and healing purposes.

Cultural Significance

3. Legal Status and Implications

The legal status of $C_{10}H_{15}N$ can lead to confusion regarding its safety and acceptability.

Evolving Legislation

Conclusion

Debunking myths and misconceptions about $C_{10}H_{15}N$ is essential for fostering informed discussions and encouraging responsible research and use. As scientific understanding of this compound continues to grow, so too must our capacity to separate fact from fiction. By addressing these misunderstandings head-on, we can pave the way for a more nuanced and informed dialogue around $C_{10}H_{15}N$, its potential benefits, and its role in both therapeutic settings and broader cultural contexts.

In the next chapter, we will explore the therapeutic uses of $C_{10}H_{15}N$ in clinical settings, examining patient outcomes and the evolving landscape of psychedelic therapy.

Chapter 17: $C_{10}H_{15}N$ in Therapy

As society grapples with increasing rates of mental health disorders, $C_{10}H_{15}N$ has emerged as a beacon of hope in therapeutic contexts. This chapter explores the evolving landscape of $C_{10}H_{15}N$'s therapeutic applications, examining current research, treatment methodologies, and patient outcomes. The aim is to provide an understanding of how this compound is being integrated into modern therapeutic practices and the implications of its use.

Therapeutic Uses

1. Mental Health Treatment

$C_{10}H_{15}N$ has gained attention for its potential in treating various mental health conditions, including depression, anxiety, PTSD, and substance use disorders.

Clinical Studies

2. Enhancing Psychotherapy

$C_{10}H_{15}N$ is not only used as a standalone treatment but also as an adjunct to psychotherapy. The compound can facilitate deeper emotional processing and help patients confront traumatic memories.

Integration with Talk Therapy

3. Addiction Treatment

Emerging research indicates that C10H15N may help in treating substance use disorders by reducing cravings and withdrawal symptoms.

Potential Mechanisms

Research Landscape

1. Ongoing Clinical Trials

Numerous clinical trials are underway to evaluate the efficacy and safety of C10H15N in various therapeutic contexts. These studies aim to gather robust data on dosage, administration methods, and long-term effects.

Key Research Institutions

2. Safety and Efficacy Studies

Ensuring the safety of C10H15N is paramount. Ongoing studies are assessing both short-term and long-term effects to better understand the risk-benefit ratio of using this compound in therapeutic settings.

Findings to Date

Patient Outcomes

1. Case Studies and Testimonials

Real-world accounts provide invaluable insights into the impact of C10H15N therapy. Many patients have reported transformative experiences, describing feelings of connectedness, clarity, and emotional release.

Personal Stories

2. Long-Term Effects

While short-term benefits are well-documented, understanding the long-term effects of C10H15N therapy is crucial for assessing its viability as a treatment option.

Follow-Up Studies

Challenges and Considerations

Despite the promising landscape, challenges remain in the integration of C10H15N into mainstream therapy. Issues such as stigma, regulatory hurdles, and the need for trained practitioners pose significant barriers.

1. Regulatory Landscape

Navigating the legal frameworks surrounding C10H15N can be complex. In many regions, its classification as a controlled substance limits research and therapeutic use.

Advocacy for Change

2. Training and Education

As the field evolves, so too must the training of healthcare providers. Understanding the complexities of $C_{10}H_{15}N$ therapy requires specialized knowledge and skills.

Development of Training Programs

Conclusion

$C_{10}H_{15}N$ holds transformative potential in the realm of therapy, particularly for mental health disorders and addiction treatment. As research continues to unveil its benefits, the importance of informed, responsible use cannot be overstated. The integration of $C_{10}H_{15}N$ into therapeutic practices represents not just a medical evolution but also a cultural shift towards understanding and embracing the complexities of mental health treatment.

In the next chapter, we will explore the economic implications of $C_{10}H_{15}N$'s therapeutic use, examining market trends and the impact on various industries.

Chapter 18: Economic Implications

The emergence of $C_{10}H_{15}N$ as a compound of interest in both therapeutic and recreational contexts is reshaping various economic landscapes. This chapter delves into the market trends surrounding $C_{10}H_{15}N$, its economic impact on related industries, and the potential for future economic development linked to its applications.

Market Trends

1. Growing Demand in the Health Sector

The recognition of $C_{10}H_{15}N$'s therapeutic potential has led to an uptick in demand from the healthcare industry. As research validates its efficacy in treating mental health disorders and addiction, pharmaceutical companies are investing heavily in developing new formulations and delivery methods.

Investment in Research

2. Shifts in Consumer Preferences

Consumer attitudes toward mental health treatment are evolving, with a growing openness to alternative therapies. This shift is driving interest in $C_{10}H_{15}N$ as a viable option, particularly among younger demographics seeking holistic approaches to mental well-being.

Market Surveys

Economic Impact on Industries

1. Pharmaceutical Industry

The pharmaceutical sector stands to gain significantly from the commercialization of $C_{10}H_{15}N$-based therapies. As clinical trials yield positive results, the potential for new medications could lead to a multi-billion-dollar market.

New Product Development

2. Healthcare Services

The integration of $C_{10}H_{15}N$ into therapeutic practices is likely to affect the healthcare services industry, including mental health clinics and rehabilitation centers.

Service Expansion

3. Emerging Markets

The growing acceptance of $C_{10}H_{15}N$ is not limited to established markets. Developing regions are beginning to explore its therapeutic applications, creating opportunities for economic growth and investment.

International Collaboration

Potential Economic Challenges

While the prospects for economic growth are promising, several challenges could impact the successful integration of C10H15N into the market.

1. Regulatory Hurdles

The regulatory landscape surrounding C10H15N remains complex and varies significantly by region. The classification of C10H15N as a controlled substance in many areas poses barriers to research, development, and commercial distribution.

Advocacy for Policy Change

2. Market Saturation Risks

As interest in C10H15N grows, there is a risk of market saturation, particularly if too many companies enter the space without differentiated offerings. This could lead to price competition that undermines the profitability of C10H15N-based products.

Need for Innovation

Future Directions

1. Innovative Business Models

As the market for C10H15N expands, new business models are likely to emerge, including subscription-based therapies and telehealth platforms that offer C10H15N-assisted treatment remotely.

Teletherapy

2. Investment Opportunities

The burgeoning interest in C10H15N is attracting investors looking to capitalize on the potential returns from this emerging market. Startups focused on innovative therapies and applications for C10H15N are increasingly seeking venture capital and private funding.

Crowdfunding and Social Impact Investing

Conclusion

The economic implications of C10H15N are significant and multifaceted. As research continues to validate its therapeutic potential, the market for C10H15N-based products is poised for substantial growth. However, the journey toward widespread acceptance and integration into healthcare systems will require navigating regulatory challenges, fostering innovation, and ensuring equitable access to therapies.

In the next chapter, we will explore innovations in delivery methods for $C_{10}H_{15}N$, examining new technologies that enhance administration and patient experience.

Chapter 19: Innovation in Delivery Methods

As interest in $C_{10}H_{15}N$ continues to grow within medical and therapeutic frameworks, advancements in delivery methods are becoming increasingly crucial. The effectiveness of any therapeutic compound is not solely determined by its chemical properties but also by how it is administered to patients. This chapter explores the innovative technologies and methods currently being developed to enhance the delivery of $C_{10}H_{15}N$, ensuring that it reaches its intended effects safely and efficiently.

New Technologies in Delivery Systems

1. Transdermal Delivery Systems

One of the most promising advancements in delivering $C_{10}H_{15}N$ is through transdermal patches. These systems allow for the gradual release of the compound through the skin, providing a controlled dosage over an extended period.

Benefits

2. Nanoformulations

Nanotechnology is making significant strides in pharmaceutical delivery systems. By creating nano-sized formulations of $C_{10}H_{15}N$, researchers are improving its solubility and bioavailability.

Mechanisms

3. Inhalation Delivery

Inhalation methods are also being explored as a way to deliver $C_{10}H_{15}N$. By using nebulizers or dry powder inhalers, $C_{10}H_{15}N$ can be absorbed directly into the bloodstream through the lungs.

Rapid Onset

Advances in Administration Techniques

1. Smart Inhalers and Dosing Devices

The advent of smart inhalers equipped with sensors is revolutionizing how medications are administered. These devices can track usage patterns, provide reminders, and even record dosing information.

Patient Engagement

2. Microdosing Techniques

Microdosing—the administration of very small doses of a substance—has garnered interest for its potential to minimize side effects while preserving therapeutic effects. Researchers are exploring how microdosing $C_{10}H_{15}N$ can be safely and effectively implemented.

Clinical Trials

3. Oral Delivery Innovations

While traditional oral dosing methods are effective, they can suffer from variability in absorption due to individual differences in metabolism and gastrointestinal conditions. New formulations are being developed to enhance the oral bioavailability of $C_{10}H_{15}N$.

Controlled-Release Formulations

Challenges and Considerations

1. Regulatory Approval

Innovative delivery methods often face stringent regulatory scrutiny. Ensuring that new technologies meet safety and efficacy standards is a critical step before they can be made widely available.

Need for Comprehensive Testing

2. Cost and Accessibility

The implementation of advanced delivery methods can raise concerns regarding cost and accessibility. Innovative technologies, particularly in the realm of nanotechnology and smart devices, can be expensive to produce and maintain.

Economic Implications

Conclusion

The future of $C_{10}H_{15}N$ is inextricably linked to the innovations in delivery methods that are currently emerging. As researchers and companies continue to develop novel technologies, the potential for $C_{10}H_{15}N$ to make a significant impact in therapeutic settings becomes even more promising. The exploration of transdermal patches, nanoformulations, and inhalation methods, among others, highlights the commitment to optimizing the administration of this revolutionary compound.

In the next chapter, we will discuss the role of education in teaching chemistry and the importance of informed discussions surrounding $C_{10}H_{15}N$, equipping future generations with the knowledge needed to navigate this complex and evolving field.

Chapter 20: The Role of Education

Education serves as the cornerstone for understanding and responsibly engaging with C10H15N, a compound that embodies both scientific innovation and cultural complexity. In this chapter, we will explore the multifaceted role of education in teaching chemistry, addressing the significance of informed discussions surrounding C10H15N, and equipping individuals with the tools to navigate its implications in society.

Teaching Chemistry: A Foundation for Understanding

1. Curriculum Development

Integrating C10H15N into chemistry curricula provides students with an opportunity to study a contemporary compound that bridges organic chemistry and real-world applications. Educators can develop curricula that not only cover the chemical properties and synthesis of C10H15N but also explore its applications, societal implications, and ethical considerations.

Interdisciplinary Approach

2. Hands-On Learning Experiences

Laboratory experiences are essential for deepening understanding. By conducting experiments related to $C_{10}H_{15}N$, students can grasp the concepts of synthesis, analysis, and application firsthand. This experiential learning can spark interest in both the scientific and cultural dimensions of chemistry.

Safety Protocols

Importance of Informed Discussions

1. Facilitating Open Dialogue

Creating a space for open dialogue about $C_{10}H_{15}N$ allows students to express their thoughts and concerns. Educators can encourage discussions around its uses, societal impacts, and the controversies surrounding it, helping students to form well-rounded views.

Critical Thinking Skills

2. Addressing Misconceptions

There are numerous misconceptions surrounding $C_{10}H_{15}N$, ranging from its effects to its legality. Education plays a crucial role in debunking these myths by providing scientifically accurate information.

Utilizing Evidence-Based Resources

Preparing Future Leaders in Science

1. Empowering Student Research

Encouraging students to engage in research related to $C_{10}H_{15}N$ can cultivate a deeper understanding of its complexities. Research projects can cover topics such as its therapeutic applications, cultural significance, and potential environmental impacts.

Mentorship Opportunities

2. Promoting Ethical Considerations

As future scientists and leaders, students must grapple with the ethical implications of their work. Education should instill a sense of responsibility regarding how compounds like $C_{10}H_{15}N$ are used in society.

Case Studies and Ethical Dilemmas

Community Engagement and Outreach

1. Public Education Initiatives

Educators can extend the conversation about $C_{10}H_{15}N$ beyond the classroom through public outreach initiatives. Workshops, community seminars, and public lectures can help demystify the compound and educate the broader community about its scientific and cultural significance.

Building Trust

2. Collaboration with Local Organizations

Partnering with local organizations can enhance educational outreach efforts. Collaborations can lead to community events, educational campaigns, and shared resources, ensuring that accurate information about $C_{10}H_{15}N$ reaches diverse audiences.

Conclusion

Education is pivotal in shaping our understanding of $C_{10}H_{15}N$ and its multifaceted implications. By fostering a comprehensive educational framework that emphasizes hands-on learning, informed discussions, and ethical considerations, we prepare individuals to engage thoughtfully with this revolutionary compound.

In the next chapter, we will explore the importance of community and support networks in providing resources for education and building supportive environments for those navigating the complexities of $C_{10}H_{15}N$ and its impacts on society.

Chapter 21: Community and Support Networks

As the discourse surrounding $C_{10}H_{15}N$ evolves, the role of community and support networks becomes increasingly vital. These networks not only facilitate education and awareness but also foster safe environments for discussion and exploration of the compound's implications. In this chapter, we will examine how communities can build supportive frameworks that empower individuals, promote informed dialogue, and provide essential resources.

Building Supportive Communities

1. Creating Safe Spaces for Discussion

One of the primary functions of community networks is to provide safe spaces where individuals can share their experiences and concerns related to $C_{10}H_{15}N$. These spaces encourage open dialogue and help demystify the compound.

Facilitated Group Meetings

2. Peer Support Programs

Peer support programs can significantly enhance community engagement. Individuals with lived experiences related to $C_{10}H_{15}N$ can offer insights and guidance to others navigating similar journeys.

Mentorship Opportunities

Resources for Education

1. Developing Informational Materials

Creating accessible educational materials about $C_{10}H_{15}N$ is crucial for community outreach. These materials should cover the compound's chemistry, potential benefits, risks, and legal status.

Fact Sheets and Brochures

2. Workshops and Seminars

Organizing workshops and seminars can provide community members with deeper insights into $C_{10}H_{15}N$. Topics can range from its chemical properties to its cultural implications and therapeutic uses.

Collaborative Events

Strengthening Connections Between Diverse Groups

1. Fostering Interdisciplinary Collaboration

Encouraging collaboration between scientists, educators, healthcare providers, and community activists can lead to more comprehensive approaches to $C_{10}H_{15}N$.

Interdisciplinary Panels

2. Engaging Cultural Representatives

Cultural organizations can play a pivotal role in addressing the societal impacts of C10H15N. Collaborating with artists, musicians, and cultural leaders can create spaces for expression and understanding.

Cultural Events

Online Communities and Resources

1. Utilizing Digital Platforms

In today's digital age, online communities can serve as powerful tools for education and support. Social media platforms and forums can facilitate connections among individuals interested in C10H15N.

Social Media Campaigns

2. Webinars and Virtual Workshops

Hosting webinars can make educational resources more accessible, especially for those unable to attend in-person events. These sessions can feature experts and allow for interactive Q&A sessions.

Recordings for Future Access

Conclusion

Community and support networks are essential in fostering an informed and engaged public around $C_{10}H_{15}N$. By creating safe spaces, providing educational resources, and promoting interdisciplinary collaboration, communities can empower individuals to navigate the complexities of this compound. As we move into the next chapter, we will explore the intersection of science and art, examining how artistic representations of $C_{10}H_{15}N$ can further enrich our understanding and appreciation of its significance.

Chapter 22: The Intersection of Science and Art

The exploration of $C_{10}H_{15}N$ extends beyond the laboratory and into the realm of creativity, where science and art converge to shape cultural narratives and societal perceptions. This chapter examines how artistic representations of $C_{10}H_{15}N$ inform, challenge, and enhance our understanding of the compound, fostering dialogue and inspiring innovation.

Artistic Representations of $C_{10}H_{15}N$

1. Visual Arts

Artists have long drawn inspiration from chemistry, using visual media to interpret and express the complexities of compounds like C10H15N. Paintings, sculptures, and installations can convey the emotional and experiential dimensions of the compound, transcending the scientific language typically associated with it.

Symbolic Interpretations

2. Literature and Poetry

Literature serves as a powerful medium for exploring the themes and implications surrounding C10H15N. Writers often delve into the psychological and emotional landscapes shaped by the compound, using narrative to illuminate the human experience.

Narrative Exploration

3. Music and Performance Art

Music and performance art have also embraced C10H15N as a source of inspiration, using sound and movement to evoke the sensations and experiences associated with its use.

- **Musical Compositions**: Artists may create pieces that reflect the emotional highs and lows of the psychedelic experience, using instrumentation and composition techniques that mimic the rhythm of the mind during altered states.
- **Performance Art**: Live performances can embody the transformative nature of $C_{10}H_{15}N$, engaging audiences in immersive experiences that challenge perceptions and provoke thought.

Collaboration Between Disciplines

1. Interdisciplinary Projects

The intersection of science and art can lead to innovative projects that bridge these fields. Collaborative initiatives often engage scientists, artists, and educators in a shared exploration of $C_{10}H_{15}N$, resulting in unique perspectives and insights.

Science-Art Exhibitions

2. Community Engagement

Artistic initiatives can serve as catalysts for community engagement, encouraging public dialogue about $C_{10}H_{15}N$ and its implications.

Workshops and Interactive Installations

The Role of Art in Shaping Perception

Art has a unique ability to shape societal perceptions and challenge preconceived notions. Through creative expression, artists can confront stigmas associated with $C_{10}H_{15}N$, offering alternative narratives that promote understanding and empathy.

1. Challenging Stigmas

Artistic representations can serve to humanize the experiences of those who engage with $C_{10}H_{15}N$, presenting their stories in ways that resonate with broader audiences. This can help dismantle stigmas and foster a more nuanced conversation about its uses and impacts.

2. Fostering Empathy and Understanding

By portraying the emotional and psychological dimensions of $C_{10}H_{15}N$, art encourages audiences to consider diverse perspectives and experiences. This fosters empathy, promoting a more compassionate dialogue surrounding the compound and its cultural significance.

Conclusion

The intersection of science and art enriches our understanding of C10H15N, allowing for a multifaceted exploration of its implications. Artistic representations serve as powerful tools for communication, fostering dialogue and challenging societal perceptions. As we move into the final chapter, we will focus on personal responsibility and informed choices, emphasizing the importance of education in navigating the complexities surrounding C10H15N and its uses.

Chapter 23: Personal Responsibility and Informed Choices

As we conclude our exploration of C10H15N, it is vital to address the themes of personal responsibility and informed choices. Understanding this compound requires more than just scientific knowledge; it demands a nuanced approach to its implications, both personally and socially. This chapter will explore the importance of education, self-awareness, and ethical considerations when engaging with C10H15N and similar substances.

Navigating Choices

1. Informed Decision-Making

In a world where information is readily accessible, the ability to make informed choices is crucial. Engaging with $C_{10}H_{15}N$—whether for therapeutic, recreational, or research purposes—requires a thorough understanding of its effects, potential benefits, and risks.

- **Research and Resources**: Individuals should seek credible sources of information, including scientific literature, educational programs, and expert opinions. Organizations focused on drug education can provide valuable insights into the complexities of $C_{10}H_{15}N$, helping individuals understand its pharmacology, legal status, and cultural implications.
- **Self-Education**: Personal research empowers individuals to make choices aligned with their values and health needs. Understanding the biochemical interactions of $C_{10}H_{15}N$, its potential therapeutic applications, and the societal narratives surrounding it can foster responsible usage.

2. Recognizing Personal Motivations

Before deciding to engage with $C_{10}H_{15}N$, it is essential to reflect on personal motivations.

- **Understanding Intent**: Whether the intent is to explore creativity, seek therapeutic relief, or participate in cultural experiences, clarity of purpose can guide individuals in making choices that are safe and beneficial. Engaging in self-reflection can help uncover underlying desires or needs that may influence the decision.
- **Potential Risks**: Recognizing the risks associated with use—such as psychological impacts, legal consequences, or social stigma—requires honest self-assessment. Individuals must weigh these risks against their motivations and desired outcomes.

Importance of Education

1. Community Education Initiatives

Educational programs aimed at various communities can foster a culture of informed decision-making surrounding $C_{10}H_{15}N$.

- **Workshops and Seminars**: Community organizations can host events that educate the public about the science of $C_{10}H_{15}N$, its cultural significance, and safe practices. These initiatives encourage open dialogue, dispelling myths and addressing concerns.
- **Peer Education**: Empowering individuals within communities to become educators can create a ripple effect. Peer-led discussions can help normalize conversations about substances, facilitating greater understanding and reducing stigma.

2. Integrating Science and Ethics

Education should encompass not only the scientific aspects of $C_{10}H_{15}N$ but also the ethical considerations surrounding its use.

- **Ethical Frameworks**: Individuals should be encouraged to engage with ethical questions, such as the implications of personal use on community health, the environment, and societal perceptions. Discussions around harm reduction, consent, and respect for cultural practices can deepen understanding of personal responsibility.
- **Critical Thinking**: Education that fosters critical thinking allows individuals to evaluate information sources, understand bias, and recognize the societal narratives that shape perceptions of C10H15N. This skill is vital for navigating the complex landscape of drug use and its implications.

Conclusion

Personal responsibility in the context of C10H15N involves a commitment to informed decision-making, self-reflection, and ethical considerations. As we continue to explore the potential of this revolutionary compound, it is essential that individuals equip themselves with the knowledge and tools needed to navigate their choices wisely.

Education serves as the foundation for these choices, empowering individuals to engage with C10H15N in a way that respects both their personal journey and the broader community context. By fostering a culture of informed dialogue and responsible action, we can embrace the complexities of C10H15N while promoting health, understanding, and innovation.

With the conclusion of this exploration, we hope readers carry forward a deeper understanding of C10H15N—not just as a chemical compound, but as a significant player in the interplay of science, culture, and personal responsibility.

Chapter 24: The Future of C10H15N

As we conclude our exploration of C10H15N, it is essential to consider its future in the realms of science, culture, and society. The ongoing research, evolving legal landscape, and shifting public perceptions surrounding C10H15N present both challenges and opportunities. This chapter will delve into potential future developments, emphasizing the importance of continued inquiry and innovation.

1. Advancements in Research

A. Therapeutic Innovations

Research on C10H15N is rapidly evolving, particularly in its therapeutic applications. Studies investigating its effects on mental health conditions such as PTSD, anxiety, and depression continue to gain momentum.

- **Clinical Trials**: Future clinical trials are likely to explore optimized dosages, delivery methods, and combination therapies. These studies aim to provide evidence-based practices that could lead to FDA approval for specific medical uses.
- **Neuroscience Discoveries**: Advances in neuroimaging techniques will enhance our understanding of how $C_{10}H_{15}N$ interacts with brain pathways. This could lead to breakthroughs in tailoring treatments for individual neurological profiles.

B. Exploring New Synthesis Methods

The synthesis of $C_{10}H_{15}N$ and its derivatives will continue to be a focus for chemists seeking to develop safer, more efficient production methods.

- **Sustainable Practices**: Innovations in green chemistry could yield more environmentally friendly synthesis routes, reducing the ecological impact associated with traditional methods.
- **Synthetic Biology**: The field of synthetic biology may offer novel pathways to produce $C_{10}H_{15}N$ through engineered organisms, providing sustainable and scalable production.

2. Changing Legal and Ethical Landscapes

The legal status of C10H15N varies globally, with many regions reconsidering their approach to this compound. The future will likely see shifts in legislation that could impact research, therapeutic use, and personal consumption.

A. Policy Reform

- **Decriminalization and Regulation**: As societal attitudes evolve, we may witness moves toward decriminalizing C10H15N and regulating its use, similar to trends observed with cannabis. Such reforms could facilitate more extensive research opportunities and safer consumption practices.
- **Harm Reduction Strategies**: Policies focusing on harm reduction—such as supervised consumption sites and educational campaigns—can help mitigate the risks associated with unsupervised use, fostering a more informed public.

B. Ethical Considerations

- **Equity in Access**: As therapies involving C10H15N become more mainstream, ensuring equitable access to these treatments will be a crucial ethical concern. This includes addressing disparities in availability based on socioeconomic status and geographic location.
- **Cultural Sensitivity**: Recognizing and respecting the cultural contexts in which C10H15N is used will be essential. Future dialogues must include diverse voices to create inclusive policies that honor traditional practices while incorporating scientific findings.

3. Cultural Shifts and Societal Impact

The narrative surrounding C10H15N is changing, influenced by cultural movements and the growing interest in alternative therapies.

A. Artistic and Cultural Expressions

- **Creative Collaborations**: Artists, musicians, and writers will continue to explore C10H15N's effects on creativity and consciousness, producing works that reflect its multifaceted nature. These cultural expressions can shape public perceptions, bridging the gap between science and art.
- **Documentary and Media**: Increased interest in documentary filmmaking and media portrayals of C10H15N's impact on society may foster broader discussions, reducing stigma and promoting understanding.

B. Community Engagement

- **Grassroots Movements**: Community-led initiatives advocating for responsible use and education about $C_{10}H_{15}N$ can empower individuals to make informed choices. These movements may drive local policy changes and create supportive networks.
- **Public Forums**: Facilitating open discussions within communities about the benefits and risks of $C_{10}H_{15}N$ will foster a more nuanced understanding, promoting dialogue rather than fear-based narratives.

Conclusion

The future of $C_{10}H_{15}N$ is poised at the intersection of scientific exploration, cultural expression, and ethical consideration. As we navigate this complex landscape, it is crucial to remain open to new findings, advocate for responsible policies, and foster an inclusive dialogue that respects diverse perspectives.

Continued education and awareness will empower individuals and communities to engage with $C_{10}H_{15}N$ thoughtfully and responsibly. As this compound evolves within our societal fabric, it presents not just a chemical entity, but a powerful lens through which we can examine our relationship with nature, science, and each other.

In closing, the journey with $C_{10}H_{15}N$ is just beginning. By embracing curiosity and fostering collaboration, we can unlock its potential for innovation, healing, and understanding in the years to come.

Chapter 25: Reflections and Moving Forward

As we reach the final chapter of this exploration into $C_{10}H_{15}N$, it is important to reflect on the multifaceted journey we have undertaken. From the fundamental principles of chemistry to the intricate dance between culture and science, this compound serves as a compelling case study of how a single substance can influence various domains of life. In this concluding chapter, we will summarize the key themes explored throughout the book, highlight critical takeaways, and consider the future trajectory of $C_{10}H_{15}N$ in our society.

1. Recap of Key Themes

A. Scientific Understanding

At its core, $C_{10}H_{15}N$ represents the intersection of organic chemistry and innovative applications. We began by examining its chemical structure and properties, recognizing how these attributes underpin its diverse uses in medicine, therapy, and beyond. The advancements in synthesis methods not only demonstrate the ingenuity of chemists but also underscore the importance of responsible scientific inquiry.

B. Cultural Significance

C10H15N's presence in pop culture, art, and literature highlights its pervasive influence on human experience. The artistic expressions inspired by this compound serve as a reminder of the profound connection between science and creativity. These cultural artifacts enrich our understanding and can facilitate important conversations about the ethical and societal implications of C10H15N.

C. Legal and Ethical Landscape

Throughout the book, we navigated the evolving legal framework surrounding C10H15N, recognizing the necessity for responsible legislation that reflects both scientific evidence and societal values. Ethical considerations around accessibility, cultural sensitivity, and harm reduction emerged as critical themes, emphasizing the need for a balanced approach in policy development.

2. Critical Takeaways

A. The Importance of Education

One of the most crucial takeaways from our exploration is the role of education in shaping public perceptions and fostering informed choices. The need for clear, evidence-based information about C10H15N and its effects cannot be overstated. Empowering individuals through education will enable them to navigate complex issues surrounding the compound more effectively.

B. Community Engagement

Building supportive communities that engage in open dialogue about $C_{10}H_{15}N$ is vital for promoting understanding and mitigating stigma. Grassroots movements and local initiatives can play a significant role in shaping societal attitudes and policies, ultimately contributing to a more informed and compassionate society.

C. Interdisciplinary Collaboration

The intersection of science, art, and culture is a rich space for exploration and innovation. Future advancements in our understanding and application of $C_{10}H_{15}N$ will benefit from collaboration across disciplines, allowing for a holistic approach that honors diverse perspectives.

3. Looking Ahead

A. Research and Development

As research on $C_{10}H_{15}N$ continues to expand, it will be essential to prioritize ethical considerations and sustainability. The scientific community must remain vigilant in addressing potential risks while exploring new therapeutic applications and delivery methods.

B. Shifting Societal Norms

The landscape surrounding C10H15N is in flux. Societal attitudes are gradually shifting toward acceptance and understanding, paving the way for potential policy reforms. This changing narrative will require continued advocacy and informed public discourse to ensure that progress is made in a responsible manner.

C. Empowerment Through Knowledge

Ultimately, the future of C10H15N lies in the hands of informed individuals and communities. By fostering a culture of inquiry, openness, and education, we can navigate the complexities of this compound and its implications, ensuring that its potential is harnessed for the greater good.

Conclusion

C10H15N is more than just a chemical compound; it is a lens through which we can examine our relationship with science, culture, and society. As we move forward, let us carry the lessons learned throughout this exploration—embracing curiosity, advocating for informed discussions, and fostering collaborative efforts across disciplines. The journey with C10H15N is ongoing, and by approaching it with an open mind and a commitment to responsible practices, we can unlock its potential to inspire, heal, and innovate.

In the end, the story of $C_{10}H_{15}N$ is a testament to humanity's quest for knowledge, understanding, and connection. As we continue to explore and engage with this revolutionary compound, let us do so with a spirit of responsibility and hope for a better future.

www.ingramcontent.com/pod-product-compliance
Lightning Source LLC
Chambersburg PA
CBHW062110220526
45471CB00010B/3680